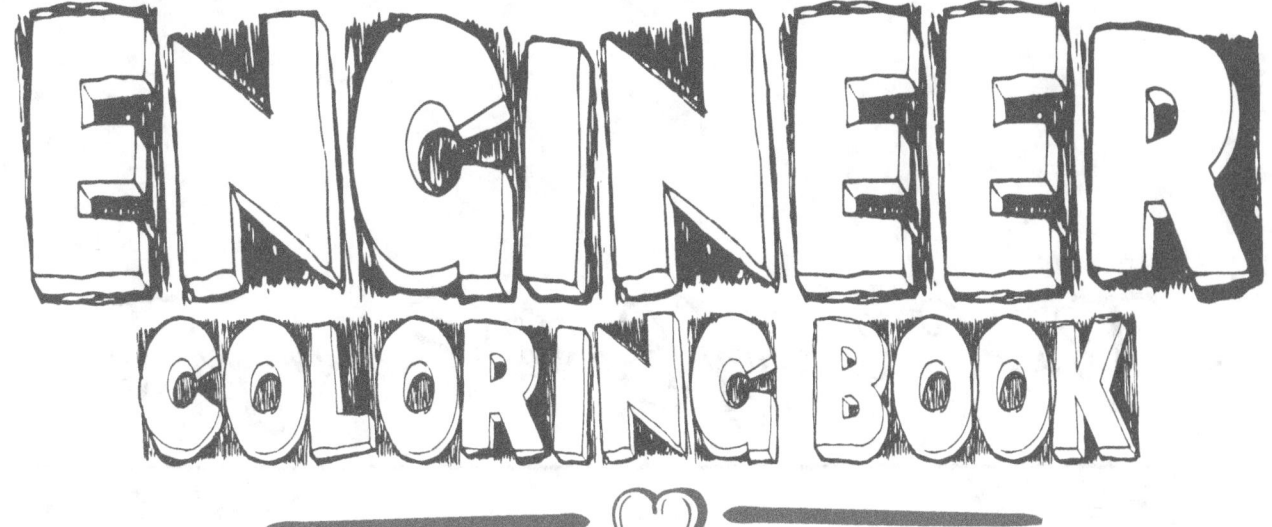

ENGINEER COLORING BOOK

A Versatile, Humorous & Motivating Adult Coloring Book for relaxing which has 28 decorative designs to relieve your stress or whenever you need a boost of confidence.

What's special?

* 15 best quote themed pages.
* 1 social media special hashtag page.
* 1 birthday special page.
* 2 decorated pages for doodling.
* 1 "show your creativity" page.
* 4 special coloring pages.
* 2 "Things i heard" coloring pages.

This Coloring Book Belongs To

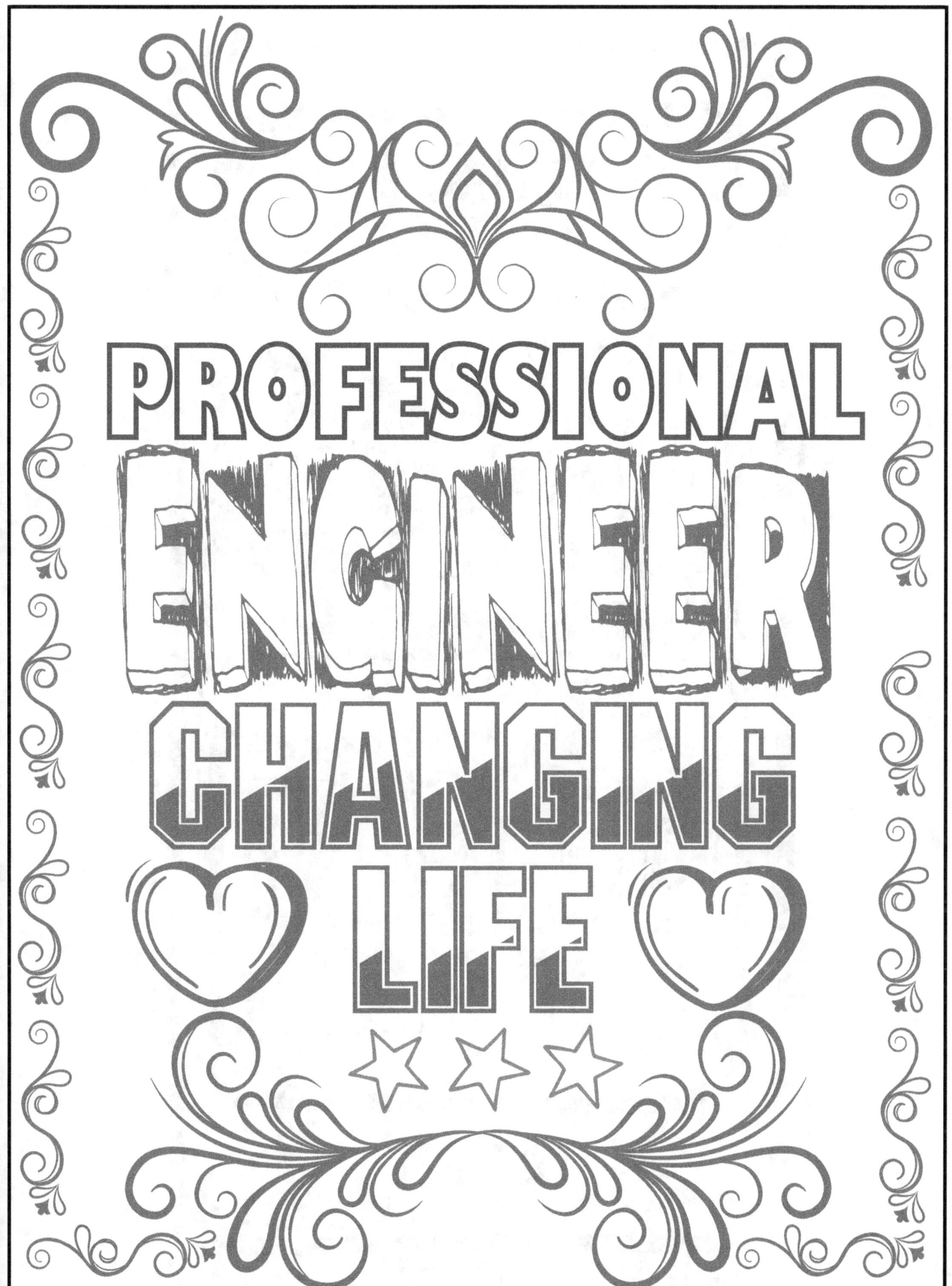

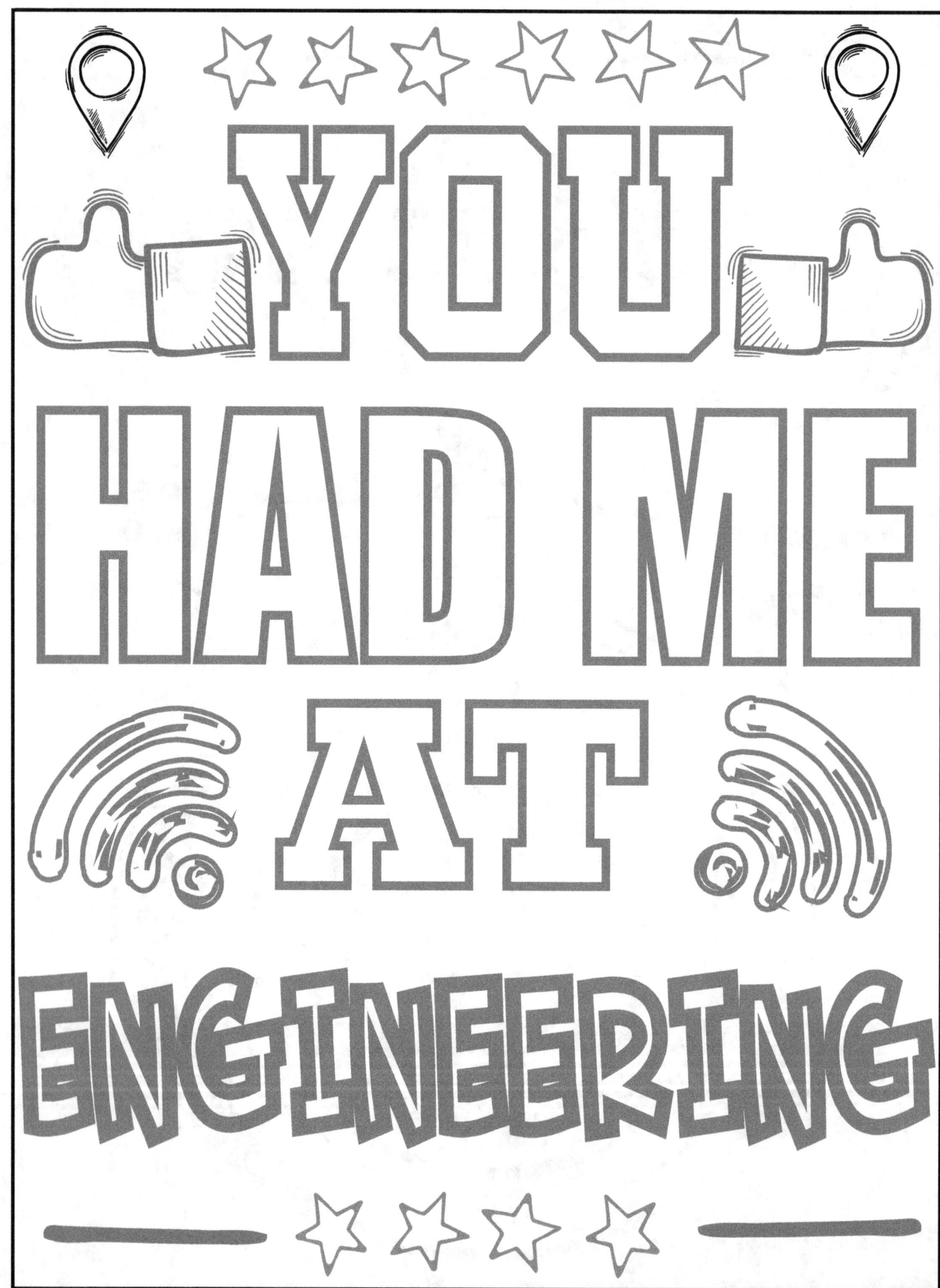

THINGS I HEARD AS AN ENGINEER THAT I WON'T FORGET

DATE : _____ / _____ / _____

HEARD IT FROM : _____

WHERE : _____

_____

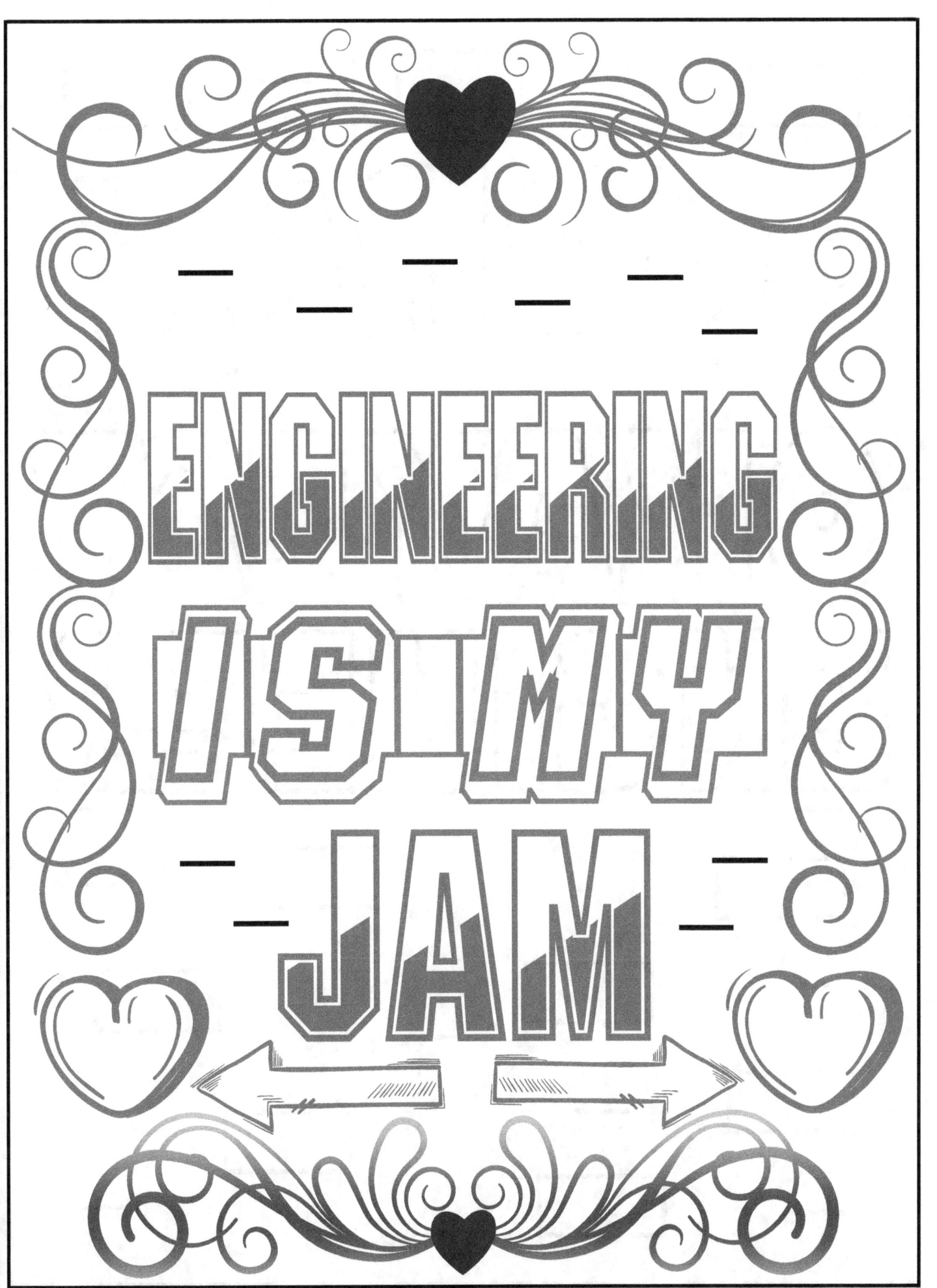

SOCIAL MEDIA

#

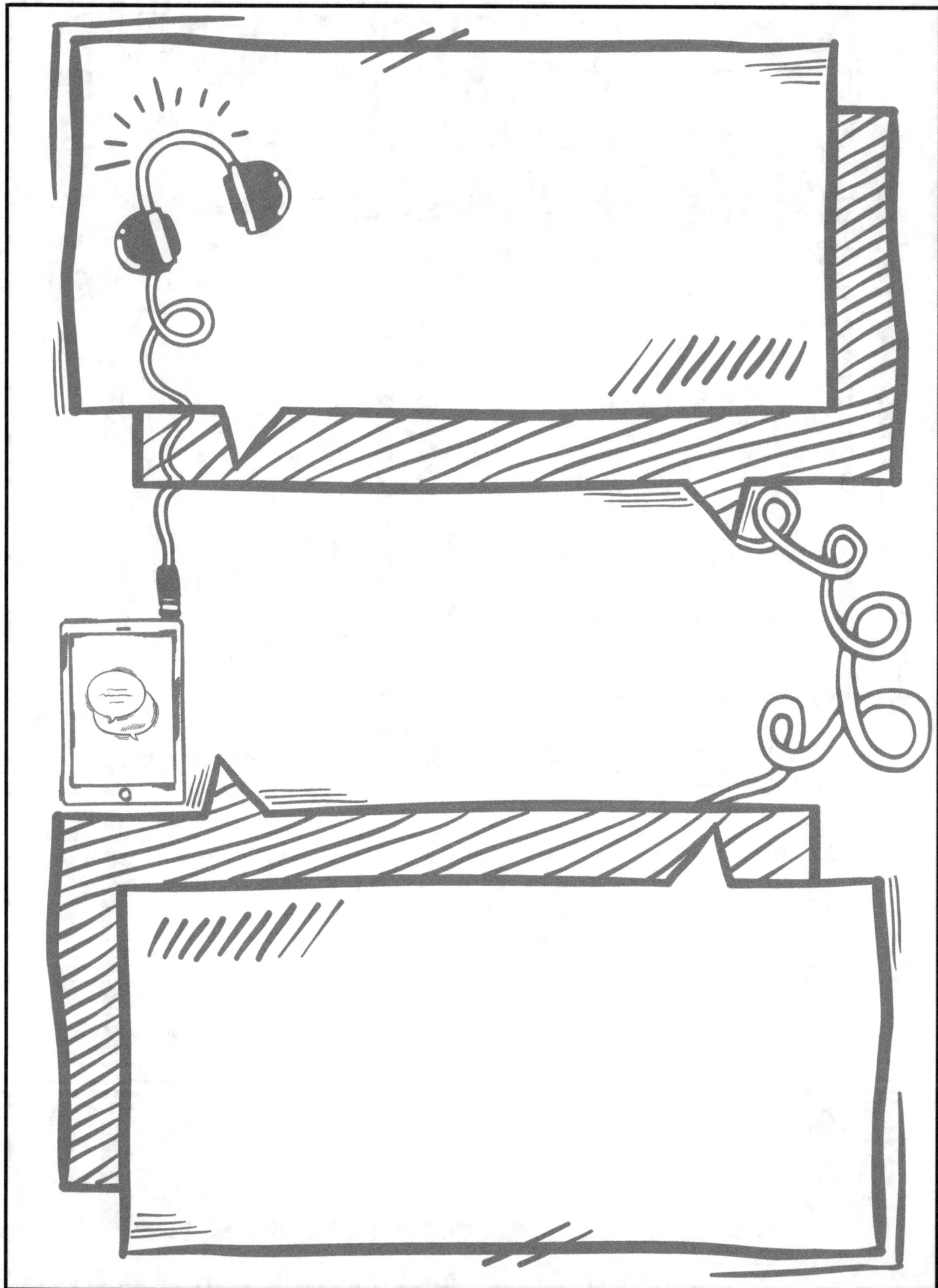